Instructions for Notebook Kee

MW01259681

1. A notebook should be kept for laboratory experiments only using a bound book. The notebook should be written in ink, and each page signed and dated. Mistakes are not to be erased but should be marked out with a single line. Try to keep your notebook with the idea that someone else must be able to read and understand what you have done. The notebook should always be up-to-date and can be collected at any time.

2. INDEX: An index containing the title of each experiment and the page number should be included at the beginning of the notebook.

3. WHAT SHOULD BE INCLUDED IN THE NOTEBOOK? Essentially everything you do in the laboratory should be in your notebook. The notebook should be organized by experiment only and should not be organized as a daily log. Start each new experiment on a new page. The top of the page should contain the title of the experiment, the date, and the page number. The page number is important for indexing, referring to previous experiments, and for labeling materials used in a given experiment. If an experiment spans more than one page, note the page on which the experiment continues if it's not on the next page. Each experiment should include the following:

   a. Title/Purpose: Every experiment should have a title and it should be descriptive. An example would be "Large-scale plasmid preparation of plasmid pXGH-5 for transfection into mouse L cells".

   b. When starting a new project, it is a good idea to introduce the overall strategy prior to beginning the first experiment. This serves two purposes. First, it forces you to think about what you are doing and why and sometimes things look differently when written down than they do in your head. Second, ideas can be patented, and a thorough description of your hypothesis and experimental strategy with appropriate documentation can be helpful for any future intellectual property issues.

   c. Many experiments should also describe the purpose of the experiment and include any information that is pertinent to the execution of the experiment or to the interpretation of the results. For example, if it is a repeat experiment, state what will be done differently to get the experiment to work. If it's a cloning experiment, include what the strategy is and how the recombinants will be screened. A simple drawing of the plasmid map can be helpful. This is not like the introduction to a paper. Include anything that will be helpful in carrying out the experiment and deciphering the experiment at a later date. For the most part, notebooks are not written for today but for the future.

4. Background information: This section should include any information that is pertinent to the execution of the experiment or to the interpretation of the results. For example, if it is a repeat experiment, state what will be done differently to get the experiment to work. If it's a cloning experiment, include what the strategy is and how the recombinants will be screened. A simple drawing of the plasmid map can be helpful. This is not like the introduction to a paper. Include anything that will be helpful in carrying out the experiment and deciphering the experiment at a later date. For the most part, notebooks are not written for today but for the future.

5. Procedure: Write down exactly what you are going to do before you do it and make sure you understand each step before you do it. In general, Xerox copies alone of procedures are not acceptable for several reasons:

    a. You should include everything you do including all volumes and amounts; many protocols are written for general use and must be adapted for a specific application.

    b. Writing a procedure out helps you to remember and to understand what it is about. It will also help you to identify steps that may be unclear or that need special attention.

    c. Some procedures can be several pages long and include more information than is necessary in a notebook. However, it is good laboratory practice to have a separate notebook containing methods that you use on a regular basis.

6. Results: This section should include all raw data, including gel photographs, printouts, colony counts, autoradiographs, etc. All lanes on gel photographs must be labeled and always identify the source and the amount of any standards. This section should also include your analyzed data, for example, transformation efficiencies, calculations of specific activities or enzyme activities.

7. Conclusions/Summary: This is one of the most important sections. You should summarize all of your results, even if they were stated elsewhere and state any conclusions you can make. If the experiment didn't work, what went wrong and what will you do the next time to try to trouble shoot?

Table of Contents

Description	Page

Table of Contents

Description	Page

Table of Contents

Description	Page

Table of Contents

Description	Page

Table of Contents

Description	Page

Table of Contents

Description	Page

Glossary

Abbreviation	Definition

Glossary

Abbreviation	Definition

Title / Experiment / Study _____

Title / Experiment / Study No. _____

Book No. _____

From Page No. _____

To Page No. _____

Witnessed & Understood by me,	Date	Invented by:
		Recorded by:

Title / Experiment / Study _____

Title / Experiment / Study No. _____

Book No. _____

From Page No. _____

To Page No. _____

Witnessed & Understood by me,	Date	Invented by:
		Recorded by:

Title / Experiment / Study _____

Title / Experiment / Study No. _____

Book No. _____

From Page No. _____

To Page No. _____

Witnessed & Understood by me,	Date	Invented by:
		Recorded by:

Title / Experiment / Study _____

Title / Experiment / Study No. _____

Book No. _____

From Page No. _____

To Page No. _____

Witnessed & Understood by me,	Date	Invented by:
		Recorded by:

Title / Experiment / Study _____

Title / Experiment / Study No. _____

Book No. _____

From Page No. _____

To Page No. _____

| Witnessed & Understood by me, | Date | Invented by: |
| | | Recorded by: |

Title / Experiment / Study _____

Title / Experiment / Study No. _____

Book No. _____

From Page No. _____

To Page No. _____

Witnessed & Understood by me,	Date	Invented by:
		Recorded by:

Title / Experiment / Study _____

Title / Experiment / Study No. _____

Book No. _____

From Page No. _____

To Page No. _____

| Witnessed & Understood by me, | Date | Invented by: |
| | | Recorded by: |

Title / Experiment / Study _____

Title / Experiment / Study No. _____

Book No. _____

From Page No. _____

To Page No. _____

Witnessed & Understood by me,	Date	Invented by:
		Recorded by:

Title / Experiment / Study _____

Title / Experiment / Study No. _____

Book No. _____

From Page No. _____

To Page No. _____

Witnessed & Understood by me,	Date	Invented by:
		Recorded by:

Title / Experiment / Study _____

Title / Experiment / Study No. _____

Book No. _____

From Page No. _____

To Page No. _____

Witnessed & Understood by me,	Date	Invented by:
		Recorded by:

Title / Experiment / Study _____

Title / Experiment / Study No. _____

Book No. _____

From Page No. _____

To Page No. _____

Witnessed & Understood by me,	Date	Invented by:
		Recorded by:

Title / Experiment / Study _____

Title / Experiment / Study No. _____

Book No. _____

From Page No. _____

To Page No. _____

| Witnessed & Understood by me, | Date | Invented by: |
| | | Recorded by: |

Title / Experiment / Study _____

Title / Experiment / Study No. _____

Book No. _____

From Page No. _____

To Page No. _____

| Witnessed & Understood by me, | Date | Invented by: |
| | | Recorded by: |

Title / Experiment / Study _____

Title / Experiment / Study No. _____

Book No. _____

From Page No. _____

To Page No. _____

| Witnessed & Understood by me, | Date | Invented by: |
| | | Recorded by: |

Title / Experiment / Study _____

Title / Experiment / Study No. _____

Book No. _____

From Page No. _____

To Page No. _____

| Witnessed & Understood by me, | Date | Invented by: |
| | | Recorded by: |

Title / Experiment / Study _____

Title / Experiment / Study No. _____

Book No. _____

From Page No. _____

To Page No. _____

Witnessed & Understood by me,	Date	Invented by:
		Recorded by:

Title / Experiment / Study _____

Title / Experiment / Study No. _____

Book No. _____

From Page No. _____

To Page No. _____

Witnessed & Understood by me,	Date	Invented by:
		Recorded by:

Title / Experiment / Study _____

Title / Experiment / Study No. _____

Book No. _____

From Page No. _____

To Page No. _____

Witnessed & Understood by me,	Date	Invented by:
		Recorded by:

Title / Experiment / Study _____

Title / Experiment / Study No. _____

Book No. _____

From Page No. _____

To Page No. _____

Witnessed & Understood by me,	Date	Invented by:
		Recorded by:

Title / Experiment / Study _____

Title / Experiment / Study No. _____

Book No. _____

From Page No. _____

To Page No. _____

Witnessed & Understood by me,	Date	Invented by:
		Recorded by:

Title / Experiment / Study _____

Title / Experiment / Study No. _____

Book No. _____

From Page No. _____

To Page No. _____

Witnessed & Understood by me,	Date	Invented by:
		Recorded by:

Title / Experiment / Study _____

Title / Experiment / Study No. _____

Book No. _____

From Page No. _____

To Page No. _____

Witnessed & Understood by me,	Date	Invented by:
		Recorded by:

Title / Experiment / Study _____

Title / Experiment / Study No. _____

Book No. _____

From Page No. _____

To Page No. _____

Witnessed & Understood by me,	Date	Invented by:
		Recorded by:

Title / Experiment / Study _____

Title / Experiment / Study No. _____

Book No. _____

From Page No. _____

To Page No. _____

Witnessed & Understood by me,	Date	Invented by:
		Recorded by:

Title / Experiment / Study _____

Title / Experiment / Study No. _____

Book No. _____

From Page No. _____

To Page No. _____

Witnessed & Understood by me,	Date	Invented by:
		Recorded by:

Title / Experiment / Study _____

Title / Experiment / Study No. _____

Book No. _____

From Page No. _____

To Page No. _____

Witnessed & Understood by me,	Date	Invented by:
		Recorded by:

Title / Experiment / Study _____

Title / Experiment / Study No. _____

Book No. _____

From Page No. _____

To Page No. _____

Witnessed & Understood by me,	Date	Invented by:
		Recorded by:

Title / Experiment / Study _____

Title / Experiment / Study No. _____

Book No. _____

From Page No. _____

To Page No. _____

| Witnessed & Understood by me, | Date | Invented by: |
| | | Recorded by: |

Title / Experiment / Study _____

Title / Experiment / Study No. _____

Book No. _____

From Page No. _____

To Page No. _____

| Witnessed & Understood by me, | Date | Invented by: |
| | | Recorded by: |

Title / Experiment / Study _____

Title / Experiment / Study No. _____

Book No. _____

From Page No. _____

To Page No. _____

Witnessed & Understood by me,	Date	Invented by:
		Recorded by:

Title / Experiment / Study _____

Title / Experiment / Study No. _____

Book No. _____

From Page No. _____

To Page No. _____

Witnessed & Understood by me,	Date	Invented by:
		Recorded by:

Title / Experiment / Study _____

Title / Experiment / Study No. _____

Book No. _____

From Page No. _____

To Page No. _____

Witnessed & Understood by me,	Date	Invented by:
		Recorded by:

Title / Experiment / Study _____

Title / Experiment / Study No. _____

Book No. _____

From Page No. _____

To Page No. _____

Witnessed & Understood by me,	Date	Invented by:
		Recorded by:

Title / Experiment / Study _____

Title / Experiment / Study No. _____

Book No. _____

From Page No. _____

To Page No. _____

| Witnessed & Understood by me, | Date | Invented by: |
| | | Recorded by: |

Title / Experiment / Study _____

Title / Experiment / Study No. _____

Book No. _____

From Page No. _____

To Page No. _____

Witnessed & Understood by me,	Date	Invented by:
		Recorded by:

Title / Experiment / Study _____

Title / Experiment / Study No. _____

Book No. _____

From Page No. _____

To Page No. _____

Witnessed & Understood by me,	Date	Invented by:
		Recorded by:

Title / Experiment / Study _____

Title / Experiment / Study No. _____

Book No. _____

From Page No. _____

To Page No. _____

Witnessed & Understood by me,	Date	Invented by:
		Recorded by:

Title / Experiment / Study _____

Title / Experiment / Study No. _____

Book No. _____

From Page No. _____

To Page No. _____

Witnessed & Understood by me,	Date	Invented by:
		Recorded by:

Title / Experiment / Study _____

Title / Experiment / Study No. _____

Book No. _____

From Page No. _____

To Page No. _____

Witnessed & Understood by me,	Date	Invented by:
		Recorded by:

Title / Experiment / Study _____

Title / Experiment / Study No. _____

Book No. _____

From Page No. _____

To Page No. _____

Witnessed & Understood by me,	Date	Invented by:
		Recorded by:

Title / Experiment / Study _____

Title / Experiment / Study No. _____

Book No. _____

From Page No. _____

To Page No. _____

Witnessed & Understood by me,	Date	Invented by:
		Recorded by:

Title / Experiment / Study _____

Title / Experiment / Study No. _____

Book No. _____

From Page No. _____

To Page No. _____

Witnessed & Understood by me,	Date	Invented by:
		Recorded by:

Title / Experiment / Study _____

Title / Experiment / Study No. _____

Book No. _____

From Page No. _____

To Page No. _____

Witnessed & Understood by me,	Date	Invented by:
		Recorded by:

Title / Experiment / Study _____

Title / Experiment / Study No. _____

Book No. _____

From Page No. _____

To Page No. _____

Witnessed & Understood by me,	Date	Invented by:
		Recorded by:

Title / Experiment / Study _____

Title / Experiment / Study No. _____

Book No. _____

From Page No. _____

To Page No. _____

Witnessed & Understood by me,	Date	Invented by:
		Recorded by:

Title / Experiment / Study _____

Title / Experiment / Study No. _____

Book No. _____

From Page No. _____

To Page No. _____

Witnessed & Understood by me,	Date	Invented by:
		Recorded by:

Title / Experiment / Study _____

Title / Experiment / Study No. _____

Book No. _____

From Page No. _____

To Page No. _____

Witnessed & Understood by me,	Date	Invented by:
		Recorded by:

Title / Experiment / Study _____

Title / Experiment / Study No. _____

Book No. _____

From Page No. _____

To Page No. _____

Witnessed & Understood by me,	Date	Invented by:
		Recorded by:

Title / Experiment / Study _____

Title / Experiment / Study No. _____

Book No. _____

From Page No. _____

To Page No. _____

Witnessed & Understood by me,	Date	Invented by:
		Recorded by:

Title / Experiment / Study _____

Title / Experiment / Study No. _____

Book No. _____

From Page No. _____

To Page No. _____

Witnessed & Understood by me,	Date	Invented by:
		Recorded by:

Title / Experiment / Study _____

Title / Experiment / Study No. _____

Book No. _____

From Page No. _____

To Page No. _____

Witnessed & Understood by me,	Date	Invented by:
		Recorded by:

Title / Experiment / Study _____

Title / Experiment / Study No. _____

Book No. _____

From Page No. _____

To Page No. _____

Witnessed & Understood by me,	Date	Invented by:
		Recorded by:

Title / Experiment / Study _____

Title / Experiment / Study No. _____

Book No. _____

From Page No. _____

To Page No. _____

Witnessed & Understood by me,	Date	Invented by:
		Recorded by:

Title / Experiment / Study _____

Title / Experiment / Study No. _____

Book No. _____

From Page No. _____

To Page No. _____

Witnessed & Understood by me,	Date	Invented by:
		Recorded by:

Title / Experiment / Study _____

Title / Experiment / Study No. _____

Book No. _____

From Page No. _____

To Page No. _____

Witnessed & Understood by me,	Date	Invented by:
		Recorded by:

Title / Experiment / Study _____

Title / Experiment / Study No. _____

Book No. _____

From Page No. _____

To Page No. _____

Witnessed & Understood by me,	Date	Invented by:
		Recorded by:

Title / Experiment / Study _____

Title / Experiment / Study No. _____

Book No. _____

From Page No. _____

To Page No. _____

Witnessed & Understood by me,	Date	Invented by:
		Recorded by:

Title / Experiment / Study _____

Title / Experiment / Study No. _____

Book No. _____

From Page No. _____

To Page No. _____

Witnessed & Understood by me,	Date	Invented by:
		Recorded by:

Title / Experiment / Study _____

Title / Experiment / Study No. _____

Book No. _____

From Page No. _____

To Page No. _____

Witnessed & Understood by me,	Date	Invented by:
		Recorded by:

Title / Experiment / Study _____

Title / Experiment / Study No. _____

Book No. _____

From Page No. _____

To Page No. _____

Witnessed & Understood by me,	Date	Invented by:
		Recorded by:

Title / Experiment / Study _____

Title / Experiment / Study No. _____

Book No. _____

From Page No. _____

To Page No. _____

| Witnessed & Understood by me, | Date | Invented by: _____ |
| | | Recorded by: _____ |

Title / Experiment / Study _____   Title / Experiment / Study No. _____

Book No. _____

From Page No. _____

To Page No. _____

| Witnessed & Understood by me, | Date | Invented by: |
| | | Recorded by: |

Title / Experiment / Study _____

Title / Experiment / Study No. _____

Book No. _____

From Page No. _____

To Page No. _____

Witnessed & Understood by me,	Date	Invented by:
		Recorded by:

Title / Experiment / Study _____

Title / Experiment / Study No. _____

Book No. _____

From Page No. _____

To Page No. _____

Witnessed & Understood by me,	Date	Invented by:
		Recorded by:

Title / Experiment / Study _____

Title / Experiment / Study No. _____

Book No. _____

From Page No. _____

To Page No. _____

Witnessed & Understood by me,	Date	Invented by:
		Recorded by:

Title / Experiment / Study _____

Title / Experiment / Study No. _____

Book No. _____

From Page No. _____

To Page No. _____

Witnessed & Understood by me,	Date	Invented by:
		Recorded by:

Title / Experiment / Study _____

Title / Experiment / Study No. _____

Book No. _____

From Page No. _____

To Page No. _____

Witnessed & Understood by me,	Date	Invented by:
		Recorded by:

Title / Experiment / Study _____

Title / Experiment / Study No. _____

Book No. _____

From Page No. _____

To Page No. _____

Witnessed & Understood by me,	Date	Invented by:
		Recorded by:

Title / Experiment / Study _____

Title / Experiment / Study No. _____

Book No. _____

From Page No. _____

To Page No. _____

Witnessed & Understood by me,	Date	Invented by:
		Recorded by:

Title / Experiment / Study _____

Title / Experiment / Study No. _____

Book No. _____

From Page No. _____

To Page No. _____

| Witnessed & Understood by me, | Date | Invented by: |
| | | Recorded by: |

Title / Experiment / Study _____

Title / Experiment / Study No. _____

Book No. _____

From Page No. _____

To Page No. _____

Witnessed & Understood by me,	Date	Invented by:
		Recorded by:

Title / Experiment / Study _____

Title / Experiment / Study No. _____

Book No. _____

From Page No. _____

To Page No. _____

| Witnessed & Understood by me, | Date | Invented by: |
| | | Recorded by: |

Title / Experiment / Study _____

Title / Experiment / Study No. _____

Book No. _____

From Page No. _____

To Page No. _____

Witnessed & Understood by me,	Date	Invented by:
		Recorded by:

Title / Experiment / Study _____

Title / Experiment / Study No. _____

Book No. _____

From Page No. _____

To Page No. _____

Witnessed & Understood by me,	Date	Invented by:
		Recorded by:

Title / Experiment / Study _____

Title / Experiment / Study No. _____

Book No. _____

From Page No. _____

To Page No. _____

Witnessed & Understood by me,	Date	Invented by:
		Recorded by:

Title / Experiment / Study _____

Title / Experiment / Study No. _____

Book No. _____

From Page No. _____

To Page No. _____

Witnessed & Understood by me,	Date	Invented by:
		Recorded by:

Title / Experiment / Study _____

Title / Experiment / Study No. _____

Book No. _____

From Page No. _____

To Page No. _____

Witnessed & Understood by me,	Date	Invented by:
		Recorded by:

Title / Experiment / Study _____

Title / Experiment / Study No. _____

Book No. _____

From Page No. _____

To Page No. _____

Witnessed & Understood by me,	Date	Invented by:
		Recorded by:

Title / Experiment / Study _____

Title / Experiment / Study No. _____

Book No. _____

From Page No. _____

To Page No. _____

Witnessed & Understood by me,	Date	Invented by:
		Recorded by:

Title / Experiment / Study _____

Title / Experiment / Study No. _____

Book No. _____

From Page No. _____

To Page No. _____

Witnessed & Understood by me,	Date	Invented by:
		Recorded by:

Title / Experiment / Study _____

Title / Experiment / Study No. _____

Book No. _____

From Page No. _____

To Page No. _____

| Witnessed & Understood by me, | Date | Invented by: |
| | | Recorded by: |

Title / Experiment / Study _____

Title / Experiment / Study No. _____

Book No. _____

From Page No. _____

To Page No. _____

Witnessed & Understood by me,	Date	Invented by:
		Recorded by:

Title / Experiment / Study _____

Title / Experiment / Study No. _____

Book No. _____

From Page No. _____

To Page No. _____

Witnessed & Understood by me,	Date	Invented by:
		Recorded by:

Title / Experiment / Study _____

Title / Experiment / Study No. _____

Book No. _____

From Page No. _____

To Page No. _____

Witnessed & Understood by me,	Date	Invented by:
		Recorded by:

Title / Experiment / Study _____

Title / Experiment / Study No. _____

Book No. _____

From Page No. _____

To Page No. _____

Witnessed & Understood by me,	Date	Invented by:
		Recorded by:

Title / Experiment / Study _____

Title / Experiment / Study No. _____

Book No. _____

From Page No. _____

To Page No. _____

Witnessed & Understood by me,	Date	Invented by:
		Recorded by:

Title / Experiment / Study _____

Title / Experiment / Study No. _____

Book No. _____

From Page No. _____

To Page No. _____

Witnessed & Understood by me,	Date	Invented by:
		Recorded by:

Title / Experiment / Study _____

Title / Experiment / Study No. _____

Book No. _____

From Page No. _____

To Page No. _____

Witnessed & Understood by me,	Date	Invented by:
		Recorded by:

Title / Experiment / Study _____

Title / Experiment / Study No. _____

Book No. _____

From Page No. _____

To Page No. _____

Witnessed & Understood by me,	Date	Invented by:
		Recorded by:

Title / Experiment / Study _____

Title / Experiment / Study No. _____

Book No. _____

From Page No. _____

To Page No. _____

Witnessed & Understood by me,	Date	Invented by:
		Recorded by:

Title / Experiment / Study _____

Title / Experiment / Study No. _____

Book No. _____

From Page No. _____

To Page No. _____

Witnessed & Understood by me,	Date	Invented by:
		Recorded by:

Title / Experiment / Study _____

Title / Experiment / Study No. _____

Book No. _____

From Page No. _____

To Page No. _____

Witnessed & Understood by me,	Date	Invented by:
		Recorded by:

Title / Experiment / Study _____

Title / Experiment / Study No. _____

Book No. _____

From Page No. _____

To Page No. _____

Witnessed & Understood by me,	Date	Invented by:
		Recorded by:

Title / Experiment / Study _____

Title / Experiment / Study No. _____

Book No. _____

From Page No. _____

To Page No. _____

Witnessed & Understood by me,	Date	Invented by:
		Recorded by:

Title / Experiment / Study _____

Title / Experiment / Study No. _____

Book No. _____

From Page No. _____

To Page No. _____

| Witnessed & Understood by me, | Date | Invented by: |
| | | Recorded by: |

Title / Experiment / Study _____

Title / Experiment / Study No. _____

Book No. _____

From Page No. _____

To Page No. _____

Witnessed & Understood by me,	Date	Invented by:
		Recorded by:

Title / Experiment / Study _____

Title / Experiment / Study No. _____

Book No. _____

From Page No. _____

To Page No. _____

Witnessed & Understood by me,	Date	Invented by:
		Recorded by:

Title / Experiment / Study _____

Title / Experiment / Study No. _____

Book No. _____

From Page No. _____

To Page No. _____

Witnessed & Understood by me,	Date	Invented by:
		Recorded by:

Title / Experiment / Study _____

Title / Experiment / Study No. _____

Book No. _____

From Page No. _____

To Page No. _____

Witnessed & Understood by me,	Date	Invented by:
		Recorded by:

Title / Experiment / Study _____   Title / Experiment / Study No. _____

Book No. _____

From Page No. _____

To Page No. _____

Witnessed & Understood by me,	Date	Invented by:
		Recorded by:

Title / Experiment / Study _____

Title / Experiment / Study No. _____

Book No. _____

From Page No. _____

To Page No. _____

Witnessed & Understood by me,	Date	Invented by:
		Recorded by:

Title / Experiment / Study _____

Title / Experiment / Study No. _____

Book No. _____

From Page No. _____

To Page No. _____

Witnessed & Understood by me,	Date	Invented by:
		Recorded by:

Title / Experiment / Study _____

Title / Experiment / Study No. _____

Book No. _____

From Page No. _____

To Page No. _____

Witnessed & Understood by me,	Date	Invented by:
		Recorded by:

Title / Experiment / Study _____

Title / Experiment / Study No. _____

Book No. _____

From Page No. _____

To Page No. _____

Witnessed & Understood by me,	Date	Invented by:
		Recorded by:

Title / Experiment / Study _____

Title / Experiment / Study No. _____

Book No. _____

From Page No. _____

To Page No. _____

Witnessed & Understood by me,	Date	Invented by:
		Recorded by:

Title / Experiment / Study _____

Title / Experiment / Study No. _____

Book No. _____

From Page No. _____

To Page No. _____

Witnessed & Understood by me,	Date	Invented by:
		Recorded by:

Title / Experiment / Study _____

Title / Experiment / Study No. _____

Book No. _____

From Page No. _____

To Page No. _____

Witnessed & Understood by me,	Date	Invented by:
		Recorded by:

Title / Experiment / Study _____

Title / Experiment / Study No. _____

Book No. _____

From Page No. _____

To Page No. _____

Witnessed & Understood by me,	Date	Invented by:
		Recorded by:

Title / Experiment / Study _____

Title / Experiment / Study No. _____

Book No. _____

From Page No. _____

To Page No. _____

Witnessed & Understood by me,	Date	Invented by:
		Recorded by:

Title / Experiment / Study _____

Title / Experiment / Study No. _____

Book No. _____

From Page No. _____

To Page No. _____

Witnessed & Understood by me,	Date	Invented by:
		Recorded by:

Title / Experiment / Study _____

Title / Experiment / Study No. _____

Book No. _____

From Page No. _____

To Page No. _____

Witnessed & Understood by me,	Date	Invented by:
		Recorded by:

Title / Experiment / Study _____

Title / Experiment / Study No. _____

Book No. _____

From Page No. _____

To Page No. _____

Witnessed & Understood by me,	Date	Invented by:
		Recorded by:

Title / Experiment / Study _____

Title / Experiment / Study No. _____

Book No. _____

From Page No. _____

To Page No. _____

Witnessed & Understood by me,	Date	Invented by:
		Recorded by:

Title / Experiment / Study _____

Title / Experiment / Study No. _____

Book No. _____

From Page No. _____

To Page No. _____

| Witnessed & Understood by me, | Date | Invented by: |
| | | Recorded by: |

Title / Experiment / Study _____

Title / Experiment / Study No. _____

Book No. _____

From Page No. _____

To Page No. _____

Witnessed & Understood by me,	Date	Invented by:
		Recorded by:

Title / Experiment / Study _____

Title / Experiment / Study No. _____

Book No. _____

From Page No. _____

To Page No. _____

Witnessed & Understood by me,	Date	Invented by:
		Recorded by:

Title / Experiment / Study _____

Title / Experiment / Study No. _____

Book No. _____

From Page No. _____

To Page No. _____

Witnessed & Understood by me,	Date	Invented by:
		Recorded by:

Title / Experiment / Study _____   Title / Experiment / Study No. _____

Book No. _____

From Page No. _____

To Page No. _____

| Witnessed & Understood by me, | Date | Invented by: |
| | | Recorded by: |

Title / Experiment / Study _____

Title / Experiment / Study No. _____

Book No. _____

From Page No. _____

To Page No. _____

Witnessed & Understood by me,	Date	Invented by:
		Recorded by:

Title / Experiment / Study _____

Title / Experiment / Study No. _____

Book No. _____

From Page No. _____

To Page No. _____

Witnessed & Understood by me,	Date	Invented by:
		Recorded by:

Made in the USA
Coppell, TX
09 January 2020